Python

Python and Javascript. The Ultimate Crash Course to Learn Python and Javascript Programming (python programming, javascript for beginners, software development)

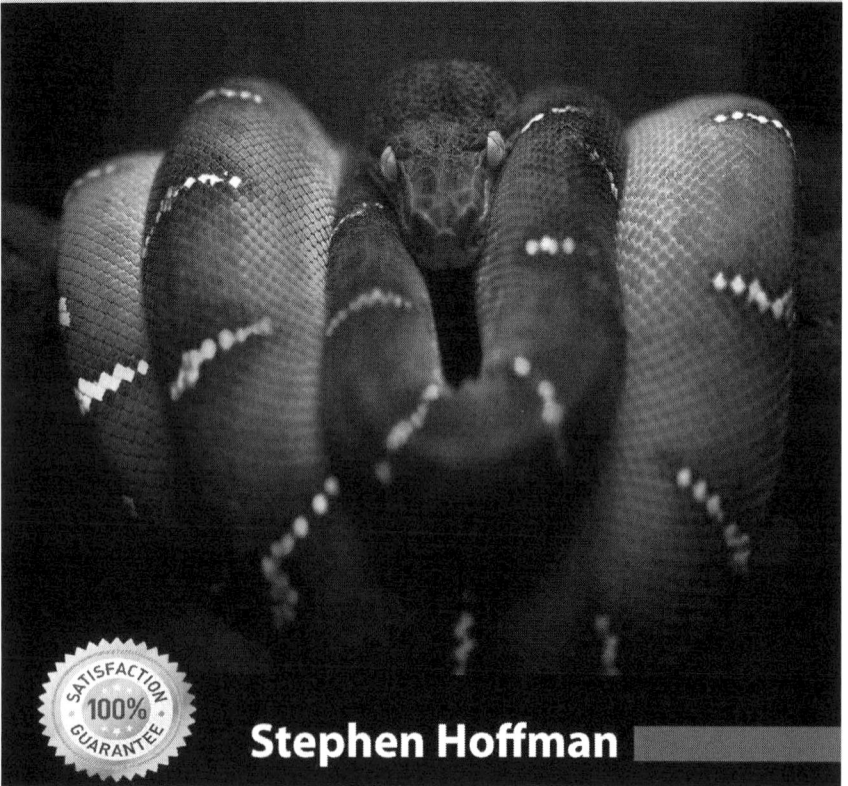

PYTHON
LEARN PYTHON FAST

The Ultimate Crash Course to Learning the Basics of the
Python Programming Language In No Time

Stephen Hoffman

Python

Learn Python FAST - The Ultimate Crash Course to Learning the Basics of the Python Programming Language In No Time

STEPHEN HOFFMAN

CONTENTS

I think next books will also be interesting for you.

C++

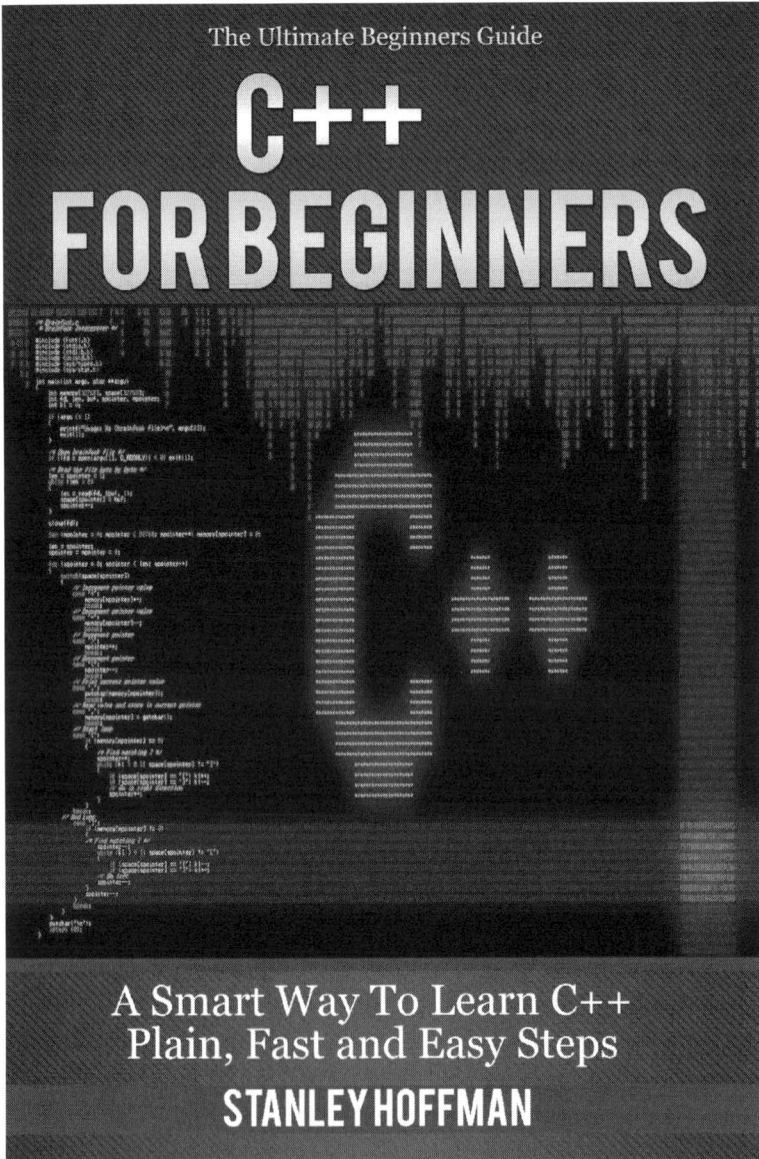

The Ultimate Beginners Guide

C++
FOR BEGINNERS

A Smart Way To Learn C++
Plain, Fast and Easy Steps
STANLEY HOFFMAN

Computer Hacking

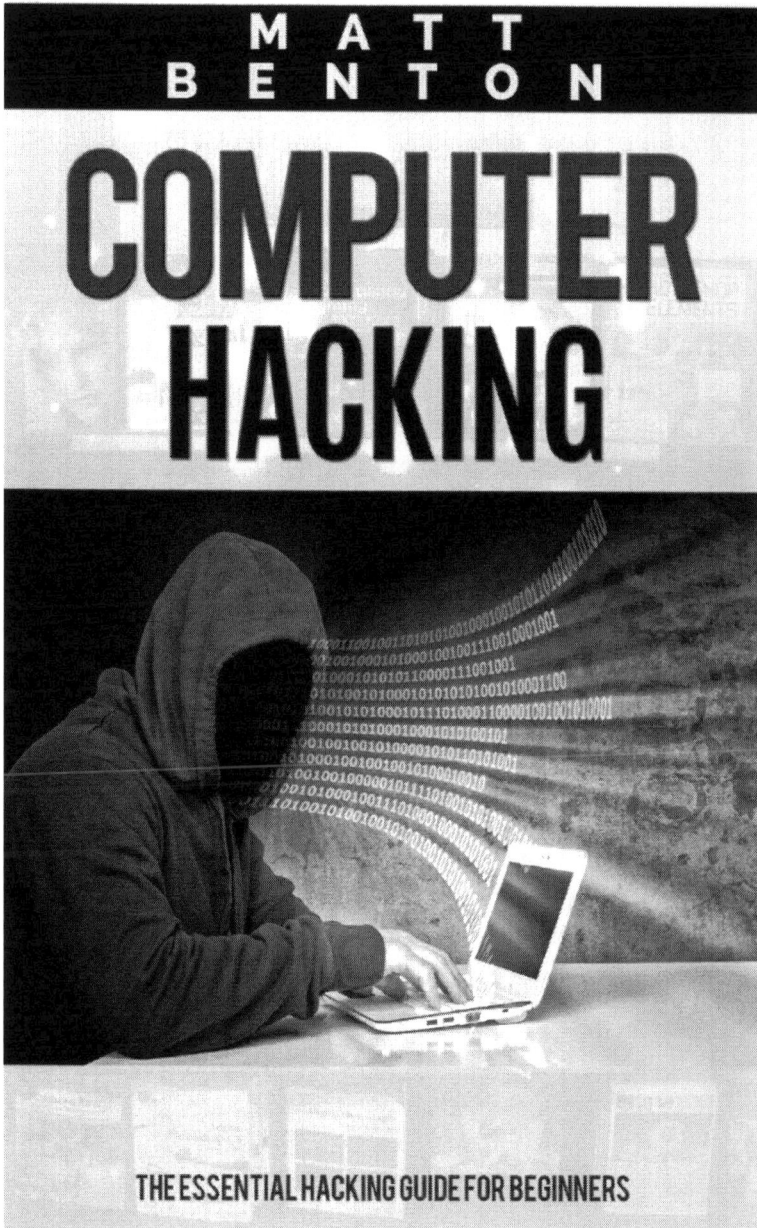

M A T T
B E N T O N

COMPUTER
HACKING

THE ESSENTIAL HACKING GUIDE FOR BEGINNERS

Hacking for Dummies

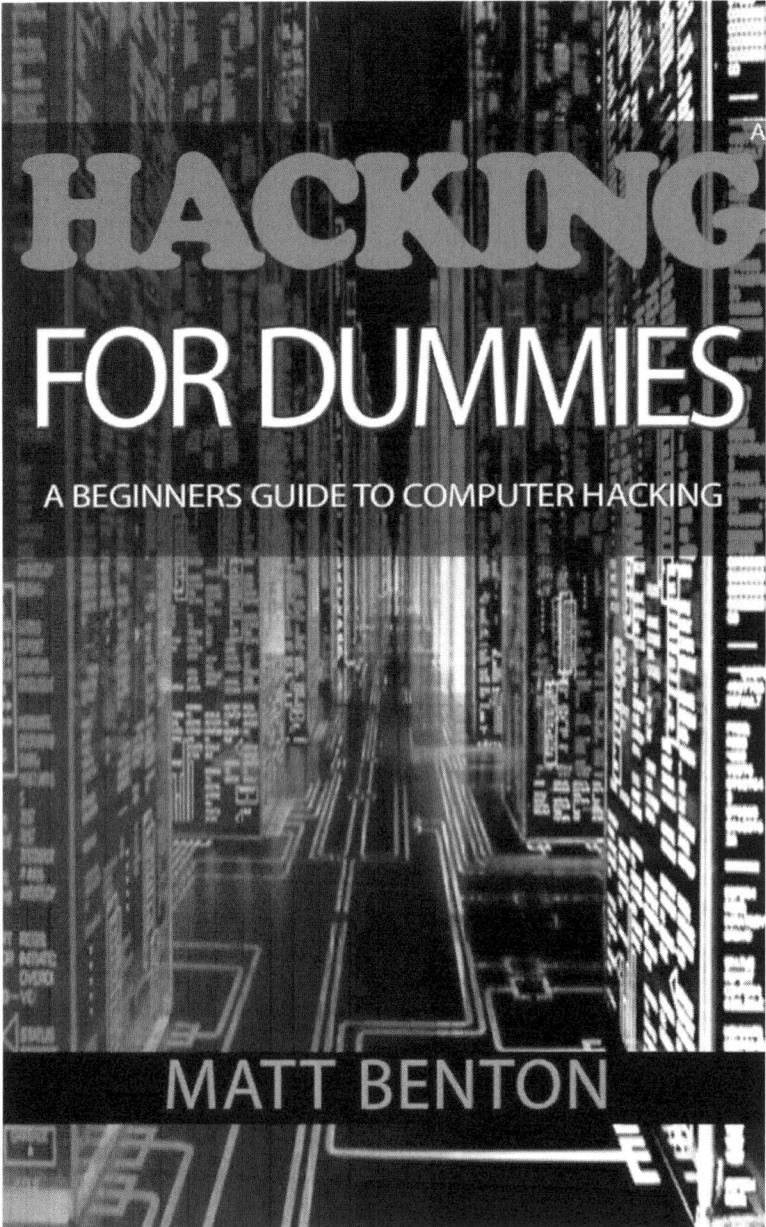

Introduction

Python can be used for a wide variety of projects, and there's a reason many programmers will reach for Python before they reach for any other programming language. Python can be used for web applications, automating tasks on systems, finding colors in images, and so many other different programming tasks!

Python is a general-purpose programming language that was created in the late 80's, and it was named after the infamous Monty Python. It's used by thousands of programmers to do anything from testing microchips at Intel to building video games with a PyGame library. It's a small language that resembles the English language and has hundreds of third-party libraries that already exist.

There are two main reasons as to why programmers will choose Python over many different programming languages.

The first reason is readability. Because it's so close to the English language, using words such as 'not' and 'in' make it so that you can read a script out loud and not feel like you're reading some long-forgotten language. This is also helped by the strict punctuation rules in Python that mean you don't have curly braces all over the code.

In addition, Python has some set rules, known as PEP 8, that tell Python developers how to format their code. This means you will always know where to put new lines and every code that you pick up from someone else will look similar and be just as easy as yours to read. That makes for some very easy collaboration between developers!

The second reason is that there are preexisting libraries for almost anything you want to do in Python. The language has been around for over twenty years, so there are many programmers out there who have mastered the language and know how to write what you want. So if you find yourself in doubt, try to find someone else who has already done it!

So the main reasons why you should learn Python are that it's simple and it's easy for you to collaborate with others in order to hone your skills. So let's get started!

Chapter One – Setting up Python

When it comes to programming in python, the first thing you should know is that python is already installed on many of the operating systems. Most operating systems, other than Windows, already come with Python installed by default. If you want to check to be sure it's installed on your computer, open the command line by running the terminal program, and type in 'python –V'.

If you see version information, then Python is already installed on your computer. However, if you get the 'bash: python: command not found' prompt, then there is a possibility that it's not installed on yours.

Follow the instructions for downloading and installing ActivePython on your computer.

Installation

The first step is to go to the Python website, www.activestate.com/Products/ActivePython, this will take you to a website where you can choose the operating system you have. If you have a Windows version that's older than Windows XP, you'll need to install the Windows Installer 2.0 before you continue.

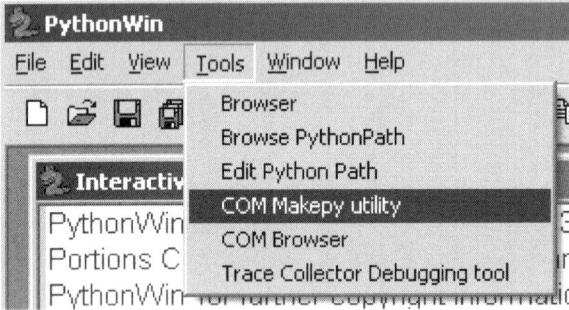

Once you find the installer you need, double click on it and go through the installation steps. After you've installed it, go to your Start-Programs-ActiveState ActivePython (versions #)-PythonWin IDE and you'll see a little script telling you what version you have and what operating system it's installed on.

Interpreter

Once you start the interactive mode with Python, it'll prompt you for the next command. This is shown with three greater than symbols, >>>. If you want to run python interpreter interactively, then you need to type in 'python' in the terminal.

It'll look something like this:

```
??$ python
Python 2.7.2 (default, Jun 20 2012, 16:23:33)
[GCC 4.2.1 Compatible Apple Clang 4.0 (tags/Apple/clang-418.0.60)] on darwin
Type "help", "copyright", "credits" or "license" for more information.
>>>

>>> 1 + 1
2
>>> you can type expressions here ... use ctrl-d to exit
```

How to Run Programs

So once Python is all set up and you know how to use the interpreter, how are you going to run programs in python? The simplest way to run a program is to type 'python helloworld.py'. If your executable is set on a .py, then the file can be run by name without having to type in the python first.

Now, set the execute bit with a 'chmod' command, such as this:
- $ chmod +x hello.py

Now you're able to run your program as ./hello.py.

Text Editor

Before you're able to write a python program in a source file, you need an editor to write that source file. When it comes to programming python, there are many editors you can choose from. Just pick one that will work with your platform. The editor you choose is going to depend on your experience with computers, what you have to do, and what platform you need in order to do it.

If you're a beginner, choose something like Notepad ++ or TextEdit.

Beginning Writing

Writing and running a python program is just a text file that is being edited by you directly. In the command line or your terminal, you can run any number of programs you want, just as you did with the aforementioned program of 'python helloworld.py'.

When the command line prompt comes up, just hit the up-arrow key in order to recall the commands that were typed previously, that way it's easy to run a previous command without retyping it.

To try out the editor you have, edit the helloworld.py program. Instead of using the word 'Hello', use the word 'Greetings'. Save the edits and run the program again to see its output. Then try adding a 'print yay!' just below the print that exists with the same indentation.

Run the program and see what the edits look like.

Congratulations! You just edited your first Python program! Now that you know how to edit a program let's take a look at variables you can use in your programs.

Chapter Two – Variables

Variables the characters you're able to use in Python. You can use any letter, every number, and every special character, but you cannot start with a number. White spaces and the signs with the special meanings like + and – are not allowed.

Either capital or lowercase letters are allowed, but most prefer to use lowercase letters in order to keep it all uniform. Variables names are case sensitive, and python is typed dynamically, so that means you don't need to declare what type every variable is. In python, the variables are a storage placeholder for texts and numbers, so it has to have a name in order to find it again.

The variable is assigned with an equal sign, followed by the value of your variable. There are a few reserved words in python that you cannot use for your variable. It's also important to know that variables can be changed later on if you need to.

For example, you can use these variables for numbers and create this formula:

- foo = 10
- bar = foo + 10

Here are a few different variable types to get you started.

# integer	X = 123
# long integer	X = 123L
# double float	X = 3.14
# string	X = "hello"
# list	X = [0,1,2]
# tuple	X = (0,1,2)
# file	X = open('hello.py', 'r')

You're also able to assign a single value to several variables at the same time with multiple assignments. Variable a, b, and c are allocated to the identical memory location with the value of 1; a = b = c = 1.

For example:

```
length = 1.10
width  = 2.20
area   = length * width
print "The area is: " , area


This will print out: The area is:  2.42
```

Now that you know how to use variables let's take a look at the interpreter in more detail.

Chapter Three – Interpreter

The interpret is usually installed at the /usr/local/bin/python location on machines where it's available. Putting /usr/local/bin in the Unix shell's search path is going to make it able to start it by typing the command: python to the shell.

When you begin the python interactive mode, it will prompt for the following command using three greater than symbols, >>>. It will then print a welcome message that states its version number and the copyright notice before it prints the first prompt.

Interactively

When used interactively, it's a good idea to have some standard comments executed each time the interpreter is begun. You can do this when you set an environment variable called PYTHONSTARTUP in the title of a file that contains start-up commands. This is like the .profile feature of the Unix shells. To do this, add this line to the .bashrc file:
>> export PYTHONSTARTUP=$HOME/.pythonstartup

Create or change the .pythonstartup file and place the python code in it, like this:

```
import os
os.system('ls -l')
```

To exit your interactive prompt, hit Ctrl+D if you're using a Linux machine.

PYTHON

Chapter Four – Importance of Comments

The comments in python are strictly for the person reading the code, whether it's you or someone else. It's a good way to communicate withgroups of people what the code is supposed to do so that they're able to collaborate better. However, there are two ways you can add comments to python that are universally accepted in the programming industry.

A single line comment is going to start with the # symbol and is terminated by the end of the line. Python will ignore all text that comes after the # symbol to the end of that line because it views that as not part of the command.

Comments that will span more than a single line is achieved by inserting a mutli-line string with """" as the delimiter on either end. They are meant as documentation for anyone who is reading the code.

Let's take a look at two examples so that you can see what I'm talking about.

```
#this is a comment in Python


print "Hello World" #This is also a comment in Python


""" This is an example of a multiline

comment that spans multiple lines

...

"""
```

So now you know how to collaborate with others and just keep your code organized with comments for yourself, let's take a look at python docstrings.

Chapter Five - Python Docstrings

Docstrings, or python documentation strings, provide a nice way to associate documentation with python functions, modules, methods, and classes. An object's docstring is made up of a string constant as the first statement in the definition of the object.

It's specified in the source code that's used to record a specific piece of code. Unlike regular source code comments, the docstring should be a description of what the function does rather than how. All functions need to have a docstring.

This lets the program inspect the comments during run time, for example, as an interactive help system or as metadata. You can use the _doc_ attribute on objects to access the docstring.

What It Should Look Like

The docstringline is going to begin with a capital letter and end with a period. The first line is going to be a short description. Don't write the name of your object. If there's more than one line in the documentation string, the second one should be blank so that it visually separates the summary from the remainder of the description. The rest of thelinesshould be one or more paragraphs that describe the object's calling conventions and anything else that's important.

An example of a docstring is as follows:

```
def my_function():
    """Do nothing, but document it.


    No, really, it doesn't do anything.
    """
    pass
```

And this is what it would like when it was printed.

```
>>> print my_function.__doc__
Do nothing, but document it.


No, really, it doesn't do anything.
```

Declaration

This python file shows you a declaration of a docstring in a python source file.

```python
"""
Assuming this is file mymodule.py, then this string, being the
first statement in the file, will become the "mymodule" module's
docstring when the file is imported.
"""

class MyClass(object):
    """The class's docstring"""

    def my_method(self):
        """The method's docstring"""

def my_function():
    """The function's docstring"""
```

Accessing the Docstring

The following session will show you how the docstring can be accessed:

```
>>> import mymodule

>>> help(mymodule)
```

Assuming this is the file mymodule.py, then the string that is the primary declaration in the file will be the mymoduleelement docstring when the file is introduced.

```
>>> help(mymodule.MyClass)

The class's docstring


>>> help(mymodule.MyClass.my_method)

The method's docstring


>>> help(mymodule.my_function)

The function's docstring
```

So now that you know what a docstring is let's talk more about some keywords in python that you need to know.

Chapter Six - Keywords in Python

Keywords in python are reserved words that are not able to be used ordinary identifiers. They have to be spelled exactly as you see them written. If you want to print the keyword list in python, then use the following commands.

```
$ python
>>>
>>> import keyword

>>> print keyword.kwlist

['and', 'as', 'assert', 'break', 'class', 'continue', 'def', 'del', 'elif', 'else',
'except', 'exec', 'finally', 'for', 'from', 'global', 'if', 'import', 'in', 'is',
'lambda', 'not', 'or', 'pass', 'print', 'raise', 'return', 'try', 'while', 'with',
'yield']
```

Definitions of Keywords

print
- o print to console

while
- o controls the movement of the platform

for

- o repeat over parts of a collection in the order they appear

break

- o break the loop cycle

continue

- o made to interrupt the present cycle without getting out of the whole cycle. A new cycle will start.

If

- o Used to figure out which statement is going to be executed

Elif

- o Is and abbreviation of else if; if the first test evaluates to false, then it will continue to the next one.

Else

- o This means the command is optional. The statement following the else word is executed unless the condition is true.

Is

- o Looks for object identity

Not

- o Disproves a Boolean value

and

- o all circumstances in a boolean expression must happen

or

- o at the minimum one condition must happen

import

- o bring in other elements into a Python script

as

- o gives a module a different alias

from

- o for introducing an exact variable, class or a function from an element def

     o makes a new user-defined function

return

     o gets out of the function and yields a value

lambda

     o makes a new unspecified function

global

     o accesses variables defined exterior to functions

try

     o states exemption handlers

except

     o holds the exception and completes codes

finally

     o is constantly completed in the end and is needed to clean up resources

raise

     o make a user-defined exclusion

del

     o deletes objects

pass

     o does nothing

assert

     o used for restoring purposes

class

     o used to create new user defined objects

exec

     o executes Python code dynamically

yield

     o is used with generators

Chapter Seven - Booleans, True or False in Python

These values are the two continuous objects of true and false. They're used to symbolize truth values, which are other values that can be considered true or false. In a numeric context, like if they're used as an argument to a mathematic operator, they will behave like the integers zero and one respectively.

Booleans and Null

True

False

None

The preexisting function bool() is able to be used to create any value to a Boolean if the value is able to be interpreted as a truthvalue. They're written as false and true.

Boolean Strings

A string can be tested for truth value. The return type is going to be in Boolean value or True or False. So let's take a look at an example. First, create a new variable and assign it a value.

```
numberone_string = "Hello World"
numberone_string.isalnum()        #check if all char are numbers
numberone_string.isalpha()        #check if all char in the string are
alphabetic
numberone_string.isdigit()        #test if string has digits
numberone_string.istitle()        #test to see if the string has title
words
numberone_string.isupper()        # test to see if the string has
upper case letters
numberone_string.islower()        #test to see if the string has lower
case letters
numberone_string.isspace()        #test to see if the string has
spaces
numberone_string.endswith('d')  #test if the string ends with a 'd'
numberone_string.startswith('H')       #test if the string begins with 'H'
```

To see what the return value would be, print it out.
numberone_string="Hello World"

```
print numberone_string.isalnum()        #false
print numberone_string.isalpha() #false
print numberone_string.isdigit()         #false
print numberone_string.istitle()         #true
print numberone_string.isupper()         #false
print numberone_string.islower()#false
print numberone_string.isspace()#false
print numberone_string.endswith('d')     #true
print numberone_string.startswith('H')   #true
```

Logical Operator Boolean

These values tend to be respond to logical operators and/or

```
>>>True and False
        False
>>>True and True
        True
>>>False and True
        False
>>>False or True
        True
>>>False or False
        False
```

Remember the built-in type of Boolean can have only one or two possible answers: True or False.

Now that you know a little more about Booleans and how to use them let's talk about python operators.

Chapter Eight - Python Operators

There are a few different types of operators. There's arithmetic, comparison, and logical operators. Let's take a look at each one in a little more detail.

Arithmetic Operators

Python has the modulus -, +,*, /, % and ** operators. Assume that the variable 'a' has a value of ten and variable 'b' has a value of twenty. Therefore:

- +
  - a+b=30
- -
  - a-b=-10
- *
  - a*b=200
- /
  - b/a=2
- %
  - b%a=0
- **
  - a**b=10^{20}
- //
  - 9//2=4

Comparison Operators

The plain comparison operators like ==, <, >=, and the others are utilized on all different manner of values. Strings, numbers, mappings, and sequences are all able to be compared. The following bullets will show you a list of comparison operators.

- < is less than
- == equal
- <= less than or equal to
- >= greater than or equal to
- > greater than
- <> not equal
- != not equal

Logical Operators

The logical operators and or also give back a Boolean value when they're used in a decision structure. There are three logical operators, or, and, and not.

For example, x>0 and x<10 is only true if x is greater than 0 and less than 10.

Operator Description:

- and: logical AND
- or: logical OR
- not: logical NOT

Now that you know about operators let's talk more about using math in python.

PYTHON

Chapter Nine - Using Math in Python

The Python language has the Python interpreter, which is a simple development environment known as IDLE, tools, libraries, and documentation. It's preinstalled on almost all Linux machines and Mac systems, but it can be an older version.

In order to begin using the calculator in the Python interpreter, type in the following script:

```
>>> 2 + 2
4

>>> 4 * 2
8

>>> 10 / 2
5

>>> 10 - 2
8
```

Counting With Variables

To put in values in the variables to count the area of a rectangle, try this code out:

```
>>> length = 2.20
>>> width = 1.10
>>> area = length * width
>>> area
2.4200000000000004
```

Counter

Counters are needed tools in programming in order to increase or decrease a value every time it is run. The following code is a counter:

```
>>> i = 0
>>> i = i + 1
>>> i
1
>>> i = 1 + 2
>>> i
3
```

Counting with a While Loop

And here's an example of when it's useful to use a counter.

```
>>> i = 0
>>> while i < 5:
...   print i
...   i = i + 1
...
0
1
2
3
4
```

The above program will count from zero to four. Between the world while and the colon, there's an expression that is going to be true at first and then become false. As long as that expression is true, the following code is going to run. The code that has to be run has to be indented. The last declaration is a counter that adds one to the value for every time the loop runs.

Multiplication Table

To make a multiplication table in Python, use the following code:

```
table = 8
start = 1
max = 10
print "-" * 20
print "The table of 8"
print "-" * 20
i = start
while i <= max:
    result = i * table
    print i, " * ", table, " =", result
    i = i + 1
print "-" * 20
print "Done counting..."
print "-" * 20
```

The output for this code is going to be as follows:
>>Output:
The table of 8
1*8=8
2*8=16
3*8=24
4*8=32
5*8=40
6*8=48
7*8=56
8*8=64
8*9=72
8*10=80
Done counting.

And that's how you use math in Python. Now let's look at exception handling in Python in the following chapter.

Chapter Ten - Exception Handling in Python

Before you get into what you can do with exceptions, you first have to know what they are. An exception is an error that occurs while a program is running. When that error happens, Python will generate an exception that can be used in order to avoid crashing the program.

Exceptions are convenient for handling special conditions and errors in programs. When you believe you have a code that is able to produce an error, then you can put in an exception handle. You can raise this exception on your own by with the raise exception declaration. Rising an exception will break the current code execution and return the exception to the previous code until it can be handled.

Here are some common errors you might see in your program

- IOError – the file is unable to be opened.
- ImportError – Python is not able to find the module.
- ValueError – this is brought up when a built-in operation or function obtains an argument that might have the right type but not the right value.
- KeyboardInterupt – This is brought up when the user will hit the interrupt key on the keyboard, this is usually the Cntrl – C or Del keys.

- EOFError – this comes up when one of the built-in functions, input() or raw_input() gets to an end-of-file conditions or EOF without scanning the data.

PYTHON

Set Up Exception Handling Blocks

In order to use exception handling in Python, first you have to have a catch-all except clause. The words 'try' and 'except' are Python keywords that programmers use in order to catch exceptions.

EXCEPTION HANDLING IN PYTHON

For example:

try — except

[exception-name] (insert exception) blocks

The code in the try clause is going to be executed statement by statement. If an exception happens, the rest of the block is going to be skipped and the except clause is going to be run.

Try:

     some statements here

Except:

     exception handling

So let's see an example of this.

try:

     print 1/0

except zerodividionerror:

     print "You can't divide by zero."

How does it work?

Error handling is handled through the use of exceptions that are made in try blocks and handled in except blocks. If your code encounters an error, a try block code execution is going to stop and transfer down to the except block.

In addition to using the except block after the try block, you're also able to use the finally block. The code in this block is going to be executed whether or not the exception happens.

Let's take a look at some code to see what will happen when you do not use error handling in a program.

Code Example

This program is going to ask the user to input a number between one and ten and then print that number.

number = int(raw_input("Enter number between 1-10"))

print "you entered number", number
This program will work just fine as long as the user enters the number, but what happens if they put in something else, like a string?
Enter a number between 1-10
hello
You can see that the program will throw an error when the string is entered.
Traceback (most recent call last):

```
File "enter_number.py", line 1, in

    number = int(raw_input("Enter number between 1-10\n"))

valueerror: invalid literal for int() with base 10: 'hello'
```

Thevalueerror is a type of exception. Now let's see how you can utilize exemption handling to repair the aforementioned program.

```
import sys

print "Let's fix the preceding code with exception handling"

try:
        number = int(raw_input("Enter a number between 1-10 \n"))

except valueerror:
        print "Only numbers are accepted."
sys.exit()

print "you entered number \n", number
```

If you now start the program and input a string in lieu of amounts, you can see that you'll get a different response. So now, let's run the program to be sure it worked. The response should be:

```
Enter number between 1-10
hello
Only numbers are accepted.
```

Try ... Except ... Else Clause

The else clause in a try, except statement has to follow all except clauses and is used for code that has to be executed if the try clause will not raise an exception. For example:

```
try:

        data = something_that_might_go_wrong
except IOError:
        handling_the_exception_error
else:
        trying_a_different_exception_handling
```

Exceptions in an else clause will not be handled by the preceding exception clauses. Be sure that the else clause is run before the final block.

Try ... Finally Clause

Thefinallyclause is an optional clause. It's used to define clean-up actions that have to be executed under all circumstances. For example:

```
try:
        raise KeyboardInterrupt
finally:
        print 'Goodbye!'
...
        Goodbye!
        KeyboardInterrupt
```

A final clause will be executed before leaving the try statement, regardless of whether an exception has happened.

Remember that if you don't specify the exception type on the exception line, it's going to catch all exceptions. This is not a good thing because it means the program will ignore any unexpected errors, as well as errors the exception block is able to handle.

Now that you know about exceptions in Python let's take a look at Strings Built-In Methods in the following chapter.

Chapter Eleven - Strings Built-In Methods

This chapter is just going to contain a list of built-in methods you can use for each string. Let's first make a string with the value of 'Hello'.

It'll look a lot like this:

string = 'Hello'

In order to manipulate this string, take a look at some of the following built-in methods and their explanations.

- string.upper()
  - This makes all the letters in the string uppercase, so if the string were 'hello jane', it would come out as 'Hello Jane'.
- string.lower()
  - This will make all the letters in your string lowercase, so if the string reads 'Hello Jane', it will come out as 'hello jane' in the program.
- string.capitalize()
  - This will capitalize only the first letter. Therefore, if the string was 'hello jane', it would read as 'Hello jane' in the program.
- string.title()

- o This would capitalize the first letter of the words. So if the string were 'hello jane' in the program, it would read 'Hello Jane'.
- string.swapcase()
    - o This string will convert all uppercase letters to lowercase and vice versa, so the following string, 'hello jane' would read 'HELLO JANE' in the program.
- string.strip()
    - o This string removes all white space from the string. So if the string was 'Hello Jane', it would read 'HelloJane' in the program.
- string.lstrip()
    - o This string removes all whitespace from the left. Therefore, if the string were ' Hello Jane ' with space before hello and one after Jane, then the code would spit out 'HelloJane'.
- string.rstrip()
    - o This string will remove all whitespace from the right. Therefore, if the same line of code were to read 'Hello Jane ' with a space after Jane, the space after Jane would be removed first to read 'HelloJane'.
- string.split()
    - o This command will split words. So if the code read 'HelloJane', then it'll be split into 'Hello Jane'.
- string.split(',')
    - o This means the code will split words by a comma, so the following string 'Hello Jane' would come out 'Hello, Jane'.
- string.count('l')
    - o This string will count the amount of times l, or any other letter you choose to put into the string, occur in the string. So the input would be 'Hello Jane' and the program would tell you 2.

- string.find(Ja)
  - o This string will find the word 'Ja' in the string.
- string.index("Ja")
  - o This will find the letters Ja in the string.
- ":".join(string)
  - o This will add a : between every character. So 'Hello Jane' would come out 'H:e:l:l:o:J:a:n:e'
- "".join(string)
  - o This will add white space between every character, so 'Hello Jane' would read 'H e l l o J a n e'
- len(string)
  - o This will find the length of the string in characters, so if the string is 'Hello Jane' then the program would say ten.

So now that you know some of the basic strings that are built-in to the Python code, let's take a look at list examples in the following chapter.

Chapter Twelve – Lists

Lists are created with square brackets, [], and the elements that are between those brackets are what the program will use. The elements in a list do not have to be the same type, ergo, you can have letters, numbers, and strings between just one set of brackets.

Let's take a look at an example.

```
myList = [1,2,3,5,8,2,5.2]

i = 0

while i < len(myList):

    print myList[i]

    i = i + 1
```

So what does this list do? The script is going to create a list, labeled (list1), with the values of 1, 2, 3, 5, 8, 2, 5, 2. The while loop is going to print each part of the list. Each part of the list list1 is going to be obtained by the index or the letter in the square brackets.

The len function will be used to go through the entire length of the list. And the final line tells the program to increase each variable by one for every time the loop runs.

So the output for this program is going to be: 1 2 3 5 8 2 5.2

Let's take a look at another example.

This example is going to count the average value of the elements that are in the list.

```
list1 = [1,2,3,5,8,2,5.2]

total = 0

i = 0

while i < len(list1):
    total = total + list1[i]

    i = i + 1


average = total / len(list1)

print average
```

The output for this will be:
#Output >> 3.74285714286

As you can see, lists are pretty easy to create and manipulate. Let's take a look at using dictionaries in Python.

Chapter Thirteen - How to Use Dictionaries in Python

You can use the curly brackets, {}, to make a dictionary in Python. You can use the square brackets, [], to index the dictionary. Separate the key and the value with some colons and with commas between the pairs. Keys have to be quoted just like with lists, and you can print out the dictionary by printing the reference to the dictionary.

A dictionary will map a set of objects (keys) to another set of objects (values). A python dictionary will map unique keys to values. Dictionaries are changeable or mutable in python. The values the keys point to are able to be any Python value. They are unordered, so the order the keys are added does not reflect with the order they might be reported back as.

Let's take a look at how to create a new dictionary in python.

In order to make a dictionary, you first have to start with an empty one. Use:
```
>>>mydict={}#
```

Example 1 – Creating a New Dictionary

This creates a dictionary that has six key-value pairs to begin with, where iPhone* is the key and the years the values.

```
released = {

                "iphone" : 2007,

                "iphone 3G" : 2008,

                "iphone 3GS" : 2009,

                "iphone 4" : 2010,

                "iphone 4S" : 2011,

                "iphone 5" : 2012

        }
print released
```

The output for this would be:
>>Output

{'iphone 3G': 2008, 'iphone 4S': 2011, 'iphone 3GS': 2009, '

iphone': 2007, 'iphone 5': 2012, 'iphone 4': 2010}

Now let's add a value to your dictionary.

```
#the syntax is: mydict[key] = "value"

released["iphone 5S"] = 2013

print released
```

>>Output

{'iphone 5S': 2013, 'iphone 3G': 2008, 'iphone 4S': 2011, 'iphone 3GS': 2009,

'iphone': 2007, 'iphone 5': 2012, 'iphone 4': 2010}

Now that you know how to do that let's remove a key and the value from the code using the del operator.

```
del released["iphone"]

print released
```

The output would be:

```
>>output

{'iphone 3G': 2008, 'iphone 4S': 2011, 'iphone 3GS': 2009, 'iphone 5': 2012,

'iphone 4': 2010}
```

Now let's check the length of the function by using the len() function.

```
print len(released)
```

Now let's test out the dictionary! Check to see if a key exists in a given dictionary by using an operator such as this:

```
>>> my_dict = {'a' : 'one', 'b' : 'two'}

>>> 'a' in my_dict

True

>>> 'b' in my_dict

True

>>> 'c' in my_dict

False
```

or this if you have a loop:

```
for item in released:
    if "iphone 5" in released:
        print "Key found"
        break
    else:
        print "No keys found"
```

or this if you need a value of a specified key:

```
print released.get("iphone 3G", "none")
```

or this if you need to print all key with a for loop:

```
print "-" * 10
print "iphone releases so far: "
print "-" * 10
for release in released:
    print release
```

or this if you want to print all the key and values:

```
for key,val in released.items():

    print key, "=>",val
```

or this if you need to get only the keys from your dictionary:

```
phones = released.keys()

print phones
```

Now that you know how to do those, why not look at how to print the values?

```
print "Values:\n",
for year in released:
    releases= released[year]
    print releases
```

Here's how to sort your dictionary:

```
for key, value in sorted(released.items()):
    print key, value
```

And that's how you use a dictionary in Python!

Conclusion

There's a lot that you can do with Python code. Many developers will use it to do something as simple as sorting some inventory to something complex like making a video game.

The important part about Python is that there are user groups all over the globes, known as PUGs, who do major conferences on continents across the globe so that Python users can interact with one another. Many of these conferences allow children to attend, who will learn how to use Python on their new Raspberry Pi's they're given at the conference.

Overall, the community surrounding the language of Python is very positive and upbeat. They are a group of people who want to make the world a better place with their knowledge. So never be afraid to reach out to forums and discussion groups to figure out how to write the code you need.

JAVASCRIPT

Ultimate Guide For Javascript Programming

STANLEY HOFFMAN

Javascript
The Ultimate guide for javascript
programming (javascript for beginners,
how to program, software development,
basic javascript, browsers)

STANLEY HOFFMAN

CONTENTS

Introduction

Computers, in this generation, seems to be no longer intimidating. They are now becoming one of our everyday tools – allowing us to make things easier and efficient. Although the latest computer operations today are user-friendly, there are still a lot of hidden complexities. Most people will find them difficult to understand. Hence, this book was written for those who want to discover the maximum potentials of computers in the simplest way possible.

In order to do that, you have to keep in mind that communication is a vital part of computer processes. This is not a process between computers but rather between the user behind the keyboard and the operating machine. The basic element, of course, is language. Learning a language like C will allow you to command the machine in the way you want and how you want it to be.

In this book, you will learn one of the most important languages of computers – the JavaScript.

Programming in a Nutshell

Before moving on to JavaScript, it is vital to understand what is programming first.

A computer program can be anything. The software you're using to read this eBook is a program. The internet browser you use is a program. The application which plays your music files is a program. The game you are playing is a program. In other words, everything you see and you use in your computer is a program. All these were products of a singleprocess known as programming.

Programming is an art. It is where programmers type a piece of text that commands the computer to perform specific operations.Let us take the human body as an analogy.

If you are given a mathematical problem to solve, your brain has to process it. Depending on the scale of the equation, you might find yourself counting your fingers or solving a formula on a sheet of paper. In short, there are several bodily functions at work.

In programming's perspective, all these body functions are programs and your whole body is the machine. The mechanical counterpart of your brain is the machine's memory. The equation is the data we want to process.

Our primary goal here is to process the data. In the scenario above, that is to have the correct answer for the mathematical problem. For our body to be able to do that, your eyes need to see the problem and relay thatto your brain. For a machine, you need a program to relay that signal to the computer's memory.

Going back to the human body, you may need to use your fingers or a sheet of paper to solve the equation. To control your fingers, your brain has to send signals for them to operate in a manner that will help you solve the problem at hand. For a machine, you need a separate program to do that.

In other words, a machine and its parts are the irrational versions of a human body and its parts. It is only that a machine can process everything in a split of a second.And to link these parts into the computer's memory, you need to assert thousands of programs. Hence, programming is the art where we create life within a lifeless body.

By now, you may find the concept of programming easier to grasp. However, the process itself is complex that even programmers get lost in their own creations.

Programs are beautiful when correctly written. Ironically, badly written programs are terrifying.

Introducing JavaScript

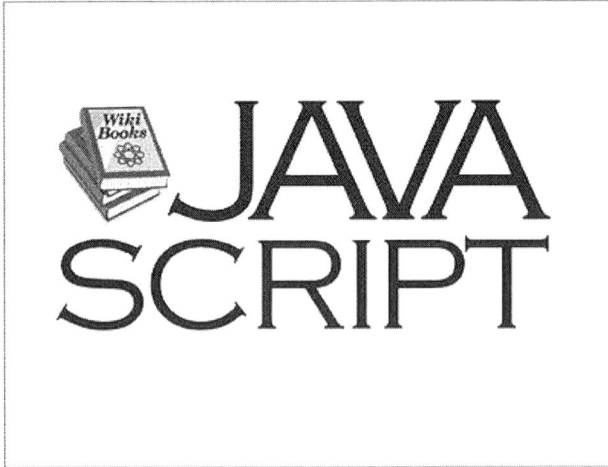

Before anything else, you might have been familiar with another programming language called Java. Keep in mind that JavaScript and Java have almost nothing to do with each other.

JavaScript was originally named as Mocha and was created in May 1995. It was the product of 10 days of hard work of Brendan Eich who was, at that time, working at Netscape. The name Mocha was picked by the Netscape's founder, Marc Andreessen.

In September of the same year, the name of the language was changed to LiveScript. In December of the same year, its name changed once again to JavaScript after receiving a trademark license from Sun. Apparently, the name was a marketing move since the Java language was being heavily campaigned and becoming more popular during those times.

Today, JavaScript is one of the programming languages that every beginner can easily learn. It is also a must-learn language for web developers alongside CSS and HTML. Web developers prefer HTML in defining their web content. They use CSS in specifying their web layouts. Lastly, they use JavaScript in programming their website's behavior. And so, JavaScript is considered to be the programming language of the web.

Even though the idea behind JavaScript was to give easier programming realm for beginners, the flexibilities offered by the language has major advantages to veteran programmers as well. It gives you numerous spaces for programming techniques that are impossible to do in more rigid programming languages. You will see more of these techniques later on this book. So if you are not new in programming but just starting to learn JavaScript, don't hate it if it usually interprets the codes you give differently. You just need to be patient and be open because JavaScript is not only fun, but it will bring you to a whole new world of programming.

What Can You Find In This Book?

This book is the first part of JavaScript in Simple Words series. This part of the series features 3 chapters. The first chapter will introduce you to values and how to process these values in JavaScript language. You will learn how to use operators, comparators, and how to create strings. The second chapter discusses variables. In this chapter, you will learn how to assign values inside variables. You will also learn how to create versatile messages by combining multiple variables. The last chapter discusses the creation of JavaScript source files. In this chapter, you will learn how to write JavaScript codes inside the source file properly. In addition, you will learn how to execute this file inside an HTML file.

The JavaScript in Simple Words book series were divided into several parts to help you master the programming language without being overwhelmed. Many beginners quitted because they thought that JavaScript is too much to handle. The entire series aims to bring you right to the heart of JavaScript in the simplest way possible.

Chapter 1 – Getting to Know Numbers, Operators, and Strings

Browsing the internet alone gives you access to millions of lines of JavaScript codes. The moment you start your favorite internet browser (e.g. Google Chrome, Mozilla Firefox, and Internet Explorer) and began to visit a certain website, you are already communicating using the JavaScript language.

In this chapter, we will be discussing the fundamental building blocks of JavaScript language. This will help you establish a concrete foundation once you move on to advanced frameworks like jQuery.

You will learn more about numbers, operators, and strings. This chapter will also cover the different processes for you to manipulate values using operators and how can you store these values inside variables. Lastly, you will also learn how to create, run, and find JavaScript codes within HTML codes.

The Console

Before anything else, it is vital for you to have access to the web browser console. The web browser console serves as one of the developer's tools. It logs information associated with the current web page. The information includes CSS, security errors, and JavaScript. The console will also allow you to engage with the web page by asserting JavaScript expressions.

This is where we will be executing our sample JavaScript codes within the book. You can access it by pressing F12 key on your keyboard. Look for the Console tab and you are now ready to begin.

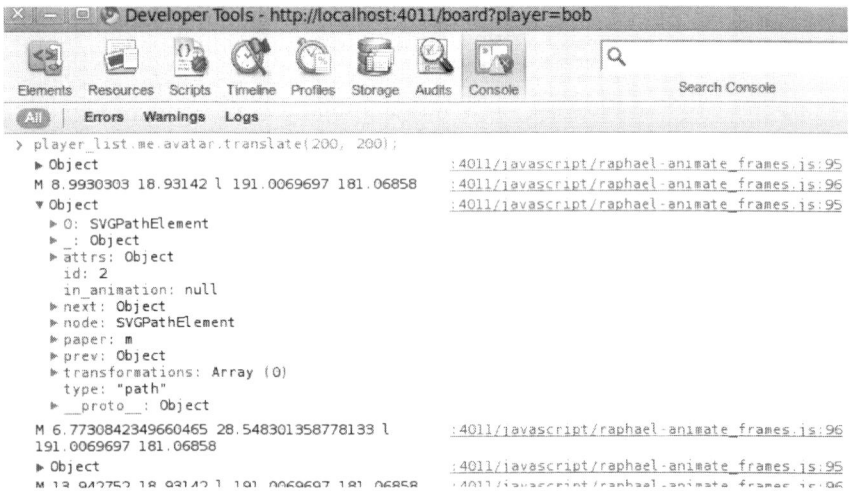

Basic Numbers and Prompt

There are 2 symbols you need to familiarize as you begin to learn JavaScript:

>   The greater than symbol; and,
→   The arrow symbol

The > symbol represents the JavaScript prompt. All the codes we want to enter into the console will be written after this symbol. The → symbol refers to the result produced by the code we entered.

In JavaScript language, numbers are automatically recognized. Writing the number after the prompt and hitting the Enter key on your keyboard will give you the same value as a result. Try to enter number 13 into the console and it will show the same value in return.

```
> 13
→ 13
```

The same goes with numbers that have decimal values.

```
> 13.29
→ 13.29
```

Operators

In JavaScript, mathematical operations are also acceptable. The operations commonly used in JavaScript syntax are the following:

- Addition
- Subtraction
- Multiplication
- Division
- Modulus

Addition is expressed through the + sign.

```
> 1 + 3
→ 4
```

Subtraction is expressed through the – sign.

```
> 10 – 5
→ 5
```

Multiplication is expressed through the * sign.

```
> 2 * 9
→ 18
```

Division is expressed through the / sign.

```
> 25 / 5
→ 5
```

Modulus, on the other hand, is a special operation in JavaScript language. It is expressed through the % sign. The result you'll get is the remainder between the two numbers after dividing them.

```
> 23 % 2
→ 1
```

In the modulus example above, the console divides 23 by 2. Even if it goes 11 times in the number 23, the console will only show the remainder as the result which is 1.

PEMDAS Order

JavaScript programming follows the popular mathematical order of operations: the PEMDAS. This acronym stands for:

P	-	Parentheses
E	-	Exponents
M	-	Multiplication
D	-	Division
A	-	Addition
S	-	Subtraction

Let us take the following expression for example.

```
> (10 + 5) * 3
```

Upon pressing Enter, the console will give you this answer in return.

```
→ 45
```

The operation inside the parentheses will be solved first. In our case, it is 10 + 5 which gives us 15. After which, the sum will then be multiplied to 3. Hence, the console gives you 45.

Let us try a longer expression.

```
> (7 * 2) + 5 – 10 / 2
```

Following the order of PEMDAS, the expression will be solved as follows:

$$14 + 5 - 10 / 2$$
$$14 + 5 - 5$$

$$19 - 5$$

Hence, the console will give you this answer in return.

$\rightarrow 14$

Comparators

Aside from operators, JavaScript language recognizes logical operators as well which we call comparators. The comparators commonly used in JavaScript are expressed through the following syntax:

>   Greater than
<   Less than
==   Equal
!=   Not equal
>=   Greater than or Equal
<=   Less than or Equal

Sending an expression in JavaScript using these comparators is like asking the console with a logical question. The answer that you will get from the console may either be "true" or "false". These answers are known to be as Boolean values.

Let us try to send the following expression to the console.

>10 != 9

This expression is like asking the console if 10 is not equal to 9 in JavaScript language. And so, the console will give you "true" as an answer since 10 is not numerically equal to 9.

Strings

```
1   <!DOCTYPE html>
2   <html>
3   <body>
4   <script type="text/javascript">
5   var str="Split String example";
6   document.write(str.split(" ") + "<br />");
7   document.write(str.split("") + "<br />");
8   document.write(str.split(" ",1) + "<br />");
9   document.write(str.split());
10  </script>
11  </body>
12  </html>
13
14  |
```

JavaScript handles, stores, and processes flat text as strings. To inform the console that we want to process flat texts, we need to enclose them within quotation marks. This will allow the console to recognize the texts as a string and will return the string back to us.

> "The quick brown fox jumps over the lazy dog"
→"The quick brown fox jumps over the lazy dog"

Multiple strings can be combined by using the concatenation method. We concatenate two or more strings by simply adding a + sign in between of strings. The console will combine all concatenated strings into a single string.

> "Lee" + "and" + "Haide"
→"Lee and Haide"

Did you notice that there is a space at the beginning and at the end of the string " and "? This is because concatenation does not automatically considers the + sign as space. Let us try another example using a combination of strings and numbers.

```
> "I will deliver this at" + 5 + "PM"
→"I will deliver this at5PM"
```

There are two things you should notice here. First, there is no need for quotation marks for numbers when concatenating them with strings. Second, the results we received is not how we want it to be. Hence, we should add a space at the end of the first string and at the beginning of the second string. It should go like this.

```
> "I will deliver this at " + 5 + " PM"
→"I will deliver this at 5 PM"
```

Aside from numbers, you can also concatenate strings, numbers, and expressions altogether. Take note, however, that the console automatically evaluates mathematical expressions.

```
> "The rooftop needs to be " + 15 + " feetand " + 1/4
→"The rooftop needs to be 15 feet and 0.25"
```

The fraction in our JavaScript code was automatically transformed into a value. If you need to keep the expression as it is, then you need to assert them within a string. It should be like this.

```
> "The rooftop needs to be " + 15 + " feet and 1/4"
→"The rooftop needs to be 15 feet and 1/4"
```

Formatting Strings with Special Characters

You can use special characters to format how strings look. One good example is the special character backslash (\ ) followed by the letter "t".

\t   Signals the console to advance to the next tab stop

Writing strings with this additional command and the result you will get should look like this.

> "Height:\t6 ft.\t\t\tWidth:\t26 ft."
→"Height: 6 ft.Width: 26 ft."

Adding quotation marks within a string should be writtenwith backslash as well.

\"   Signals the console that the following quotation mark is not the start or the end of a string.

Here is how your string line should look like when asserting quotation marks inside the string and the output you will get in return.

> "Nicknames:\t\"The Geek\" \"Com Wiz\" \"Backslasher\""
→"Nicknames: "The Geek" "Com Wiz" "Backslasher""

The same rule applies if you want to assert the backslash symbol in your string output.

\\   Signals the console to add the
          following backslash as it is in
          your output.

When writing this command, the string and the output should appear as follows.

> "Target Folder:\t\"C:\\Program Files\\Directory 1\\Directory 2\""
→"Target Folder: "C:\Program Files\Directory 1\Directory 2""

You can also command the console to put values to a new line by simply adding the letter "n" after the backslash.

\n   Signals the console to assert
          the next values in a new line.

This is how your code and the output should look like.

> "Member 1:\tFebz\nMember 2:\tDonz\nMember 3:\tTaz"
→"Member 1:   Febz
→ Member 2:   Donz
→ Member 3:   Taz"

Comparators and Strings

Aside from numbers, comparators can also be used to compare strings to get Boolean values as an output. Let us try to use the equal comparator.

> "Obi-Wan Kenobi" == "Obi-Wan Kenobi"
→true

Basing from the example above, we got the Boolean value "true" in return because the first string is exactly the same with the second string. Let us try to alter one of these strings.

> "Obi-Wan Kenobi" == "Obi-Wan Kinobi"
→ false

Keep in mind that JavaScript language is case sensitive. Hence, if the first string was written with uppercase characters and the second string was completely written in lowercase characters; then the console will treat them as not equal.

> "Obi-Wan Kenobi" == "obi-wan kenobi"
→ false

Counting String Length

In JavaScript language, you can easily access the length of your strings. Doing so will command the console to count all the characters including the spaces and non-alphabet characters between the quotations mark that signify the beginning and end of your string. We do it by adding the following property right after the string.

.length   Commands the console to count the characters within the string it follows.

This is how we write it and what kind of output we will get.

> "Pneumonoultramicroscopicsilicovolcanoconiosis".length
→45

Now let us try to use a string with spaces and non-alphabet characters.

> "Pink, Blue, Black, Red, Purple, White, and Green".length
→48

Chapter 2 – Discovering Variables

Now that you already know how to write, format, and process data in JavaScript language, it is now time for you to learn about variables. In this chapter, you will learn how to store values using variables. You will not be able to do it without learning how to write variables first. Without further due, let us begin our journey to the world of variables.

```
<!DOCTYPE HTML>
<HTML>
        <HEAD>
                <TITLE>Variables</TITLE>

                <SCRIPT LANGUAGE = "Javascript">

                        var number1 = 42;
                        var true_or_false = true;
                        var firstName = 'Kenny';

                        document.write(number1 + "<BR>");
                        document.write(true_or_false + "<BR>");
                        document.write(firstName + "<BR>");

                </SCRIPT>
        </HEAD>

        <BODY>

        </BODY>
</HTML>
```

The Storage of Values

In order to manage data in JavaScript language, you to store them inside variables. Let me show you a basic assignment syntax for a variable we shall name as "Learn Easy".

```
>varlearnEasy = 6
```

The code "var" here refers to the variable keyword. This commands the browser to reserve a specific space for this variable. The next one is "learnEasy" which refers to the variable name. This is the name we will be assigning to this example variable syntax. The equal sign here serves as the assignment operator. Notice that we only use a single equal sign here unlike the double equal signs that we used previously for comparison. Lastly, the number 6 here serves as the value which we are storing inside our variable named as learnEasy.

Now that we have a stored value inside a variable, we can now call it using the variable name alone. Asserting the variable name "learnEasy" in the console will give us the assigned value as its output.

```
>learnEasy
→ 6
```

When assigning names in our variables, there are rules we need to follow in JavaScript language. These rules are as follows:

     1. No spaces in the variable name. (E.g. var learn easy, var sweet java)
     2. No digits at the beginning of the variable name. (E.g. var 4life, var 1piece)

There are special characters you can use in the variable name. But, it is better to use simple names to make JavaScript languageless complicated.

      1. You can use underscores (_) in the variable name. (E.g. varlearn_easy)
      2. You can use the dollar sign ($) in the variable name. (E.g. varlearnEa$y)
      3. You can use both underscores and dollar signs together. (E.g. varlearn_ea$y)

Here is what a good variable name should be. It is known as the "Camel Case" which begins with lowercase and the first letter of the following words in uppercase.

```
>varsimpleAndClean
```

If you like to reuse a variable name to store a different value, then you can simplify everything by adding a number at the end like this.

```
>var simpleAndClean2
```

Changing Assigned Value in Variables

Let us go back with our example variable learnEasy. At the moment, we have the numerical value 3 assigned to it. Let us say, for example, you want to change the value assigned to that variable. You can easily do it by entering the variable name, followed by the assignment operator, then followed by the new value you want to assign to the variable.

```
>learnEasy = 13
```

Notice that we no longer use the variable keyword here. This is because the console remembers that the value is already existing in its memory. Once you call for this variable, you will now get the new value assigned to it.

>learnEasy
→ 13

You may also change the value of an existing variable using its own variable name in mathematical expressions. The format should be the variable name followed by the assignment operator. You insert the mathematical expression after the operator. It should go like this.

>learnEasy = learnEasy + 16

This will change our store value to 29 since the expression calls for the current variable value, which is 13, plus 16. This makes it 29.

There is another way to do this. We will end up changing the stored value using a different syntax. It goes like this.

>learnEasy += 2

The += operator here commands the console to add the current value of the variable to the number 2 first. Then it commands the console to store the new value to the same variable afterward. Since our current value is 29, we will now get 31 as the new value stored in the variable learnEasy.

All operators can be used in this case using both formats.

Storing Strings in Variables

Aside from numerical values, you can also store strings inside variables. The syntax uses the same format. Let us create a variable to welcome our guests.

```
>var hello = "Welcome to my website!"
```

By storing the string in a variable, you no longer have to type lengthy strings. Just call the name of the variable and the console will give you the stored string. This is how it should appear.

```
>hello
→"Welcome to my website!"
```

You can also use the length property using the variable name. The output value you will get is the same with adding the length property after the string. Let us take, for instance, a variable named "fox" which contains the string "The quick brown fox jumped over the lazy dog". We simply add .length after the variable name and we will have the following lines on our console:

```
>fox.length
→ 44
```

Finding Characters in Strings

The JavaScript language has its own way to index each character inside the string. Let us take the variable "hello" as an example.

```
"  W  e  l  c  o  m  e     t  o     m  y     w  e  b  s  i  t  e  !  "
          0   1  2  3   4  5   6  7  8   9 10 11   12 13 14 15 16 17
          18 19 20 21
```

When indexing, JavaScript indexes the first character using the numerical value 0. This is different from the .length property since JavaScript counts the first character using the numerical value 1. In other words, the last index number is always 1 value less than the length value.

Using the string stored in hello variable, we have 22 for length property and 21 as the last index number. Now, why do we need the index number you ask?

We use it to find specific characters within the string. This is done using the charAt( ) method which we place after the variable just like the .length property. For instance, we want to know which character falls under index number 20 from our variable hello. We simply type the following:

```
>hello.charAt(20)
→ "e"
```

Creating Messageswith Variables

Now that you know how to store values inside variables, let us now create messages using variables. Let us create one more variable to combine with our existing variable "hello".

> ```
> >var enjoy = "Please enjoy your visit!"
> ```

Now that we have our second variable, let us now combine them and create a new variable that will serve as our welcoming message.

> ```
> >varwelcome = hello + enjoy
> ```

This will give us the following output whenever we call for the variable welcome.

> ```
> > welcome
> →"Welcome to my website! Please enjoy your visit!"
> ```

Chapter 3 –Building and Running JavaScript

Learning variables was fun, was it not? However, no one wants to code JavaScript language in a console. In this chapter, you will learn how to build JavaScript files. You will also learn how to run these files inside HTML. So let us begin.

Creating Source Files

Writing JavaScript codes in the console alone will not print out what we want to convey to our user's web browsers. Hence, we need to find a way to bring what we want to the users. In order to do that, we need to create JavaScript source files and run these files in an HTML file.Here is an example of the most common HTML file known as index.html.

```
<html>
<head>
<script src="greetings.js"></script>
</head>
<body>
<h1>WELCOME TO JAVASCRIPT!</h1>
...
</body>
</html>
```

As you can see, we have the <script>tag in this file. This tag commands this HTML file to load a specific JavaScript source file so that the web browser of your visitors will let them see your website in the way you want them to see it. Notice that we have the "src" keyword inside the <script>tag. This is a command to the HTML file which directs it where to load the JavaScript source file.

In this example, we are commanding the index.html file to load all variables and values within saved inside greetings.js.

In order to create this .js file, all you need to do is open any text editor like Notepad or Text Edit. Write all your JavaScript codes inside and save the file in .js format. Inside this file, we assert all the JavaScript codes we want our visitor's web browser to execute as soon as they access our website.

Where To Place The Source File?

When browsing your website's file manager, you will never miss seeing a folder named "root/". This is where we can find everything inside our website. This is also the default location of your website's index.html as well.

Basing from the index.html file we have above, we are directing it to find the JavaScript source file within the same location folder – that is, the root/ folder. Hence, we should place our greetings.js in the same location.

```
+ root/
  - index.html
  - greetings.js
```

There are websites, however, that place all their JavaScript source files in a subfolder. The most common directory for their source files is under scripts/ folder. Here is how they look like in your website's directory.

```
+ root/
  - index.html
  + scripts/
    - greetings.js
```

If this is how you want to organize your source files, then we need to alter the <script> tag in our sample index.html. This is how your HTML codes should look like.

```
<html>
<head>
<script src="scripts/greetings.js"></script>
</head>
```

```
<body>
<h1>WELCOME TO JAVASCRIPT!</h1>
...
</body>
</html>
```

Notice that in this code, we are instructing the index.html to load the greetings.js from the scripts/ subfolder.

Writing Codes in the Source File

Writing console-style expressions inside the greetings.js as they are will only end up in error. Let us try to write these following lines inside our source file for example.

```
var hello = "Welcome to my website! "
var enjoy = "Please enjoy your visit!"
hello + enjoy
```

Once we open our index.html and go to our browser's console, we will see nothing but error. This is because web browser got confused on what you really intend to show. The error comes from our JavaScript source file. Unlike the console, we need to assert our JavaScript codes differently in the source files. We do this by adding a semicolon (;) every after a statement. Hence, we should write our JavaScript codes like this.

```
var hello = "Welcome to my website! ";
var enjoy = "Please enjoy your visit!";
hello + enjoy;
```

With our codes written with semicolons, you will no longer see an error when trying to access the console. However, you will not see anything in the console. This means that our code is not yet being delivered to the browser's console. The execution of the JavaScript codes only happens within the source file itself. In order to fix this, we need to use the console.log( ) method.

What we want to do here is to show the string coming from the concatenation of the variables hello and enjoy. Here is how we are going to that.

```
var hello = "Welcome to my website! ";
```

```
var enjoy = "Please enjoy your visit!";
console.log(hello + enjoy);
```

By using the console.log( ) method, you will be able to see now the string inside the console. In other words, we now have a working JavaScript programming language being run by our index.html file.

Conclusion

At this point, you now know how to write values in JavaScript language. You just learned how JavaScript interprets numbers and flat texts, as well as how to transform texts into string values. You also learned how to manipulate these values into expressions and statements with the use of operators and comparators.

In the second chapter, you were introduced to variables. You learned how to store values into variables and the concatenation method where you learned how to process values quickly in the form of variables. A couple of methods were also introduced such as the charAt( ) method and .length property – tools that can help you analyze values efficiently.

In the third chapter, you learned how to create your own JavaScript source file. Take note that you need to assert JavaScript codes differently in source files compared to the console. You also got to know the console.log( ) method which allows you to print values in the browser's console. However, this is not yet the most efficient way to deliver JavaScript codes. Imagine if you are going to deliver hundreds of values into the browser? That would mean repetitive programming and too much hassle.

In the next part of the JavaScript in Simple Words series, we will bring you to the loops of JavaScript where you will learn how to manage repetitive coding efficiently.Several topics will also be discussed on the second part of the book series including, but not limited to, the use of conditions and the creation of dialog boxes.

Thank you for reading. I hope you enjoy it. I ask you to leave your honest feedback.

14185512R10057

Printed in Poland
by Amazon Fulfillment
Poland Sp. z o.o., Wrocław